4TH GRADE SCIENCE

Unit 3

Table of Contents

LEADERSHIP 101
Flexibility
Willingness to change plans or ideas without getting upset

ResponsiveEd® thanks Character First (www.characterfirst.com) for permission to integrate its character resources into this Unit.

- Define ecosystem.

- Understand how organisms, populations, and communities are a part of ecosystems.

- Describe different habitats.

- Describe different biomes.

- Understand how different plant adaptations help plants survive.

- Understand how different adaptations help animals survive.

- Define producer and consumer.

- Draw a diagram of photosynthesis.

- Understand that plants need water, sunlight, and carbon dioxide to do photosynthesis.

- Understand how energy flows in a food chain and a food web.

- Understand the difference between a food chain and a food web.

- Describe predator and prey.

- Describe different adaptations that predators and prey have that help them survive.

- Describe different ways ecosystems can change.

- Describe how ecosystem change can cause some organisms to be endangered or extinct.

- Understand how fossils help scientists learn how change affects some organisms.

- Describe different ways to conserve natural ecosystems.

OBJECTIVES:

- Define ecosystem.
- Understand how organisms, populations, and communities are a part of ecosystems.
- Describe different habitats.
- Describe different biomes.
- Understand how different plant adaptations help plants survive.
- Understand how different adaptations help animals survive.

VOCABULARY:

abiotic *[ey-bahy-OT-ik]* – (adjective) nonliving

adaptation *[ad-uhp-TEY-shuhn]* – (noun) a trait that helps a plant or animal survive

biome *[BAHY-ohm]* – (noun) an ecosystem with certain plants, animals, and weather that covers a large area

biotic *[bahy-ot-ik]* – (adjective) living

community *[kuh-MYOO-ni-tee]* – (noun) all the different populations living in the same place at the same time

ecosystem *[EK-oh-sis-tuhm]* – (noun) all the living and nonliving things that interact in a specific place

habitat *[HAB-i-tat]* – (noun) the place an organism lives in its ecosystem

interact *[in-ter-AKT]* – (verb) to act in a way that affects something else

organism *[AWR-guh-niz-uhm]* – (noun) a living thing

population *[pop-yuh-LEY-shuhn]* – (noun) a group of one kind of organism living in one place at the same time

structure *[STRUHK-cher]* – (noun) a body part

1. WHAT IS AN ECOSYSTEM?

Milly the bear is thirsty. "Must ... get ... water," she says. She has traveled far looking for water. She comes across what looks like a pond. She drinks from the pond and then she hears screaming. She doesn't understand why anyone would scream. There is enough water for everyone.

Bears need certain things to survive. All the things a bear needs to survive can be found in an ecosystem. An **ecosystem** is all the living and nonliving things that **interact** in a specific place. A living thing is called an **organism**. An ecosystem can be any size, from the entire planet to a long-standing puddle of water.

Ecosystems must have certain things for organisms to be able to live there. An ecosystem must have water, certain gases, shelter, and food sources. Gases that organisms need are found in air or water.

You are part of the ecosystem where you live. Have you ever studied your backyard or nearby park to see how plants and animals interact with the environment? Let's separate parts of an ecosystem into different categories.

Nonliving Parts of an Ecosystem

Nonliving parts of an ecosystem are called **abiotic**. Here are some abiotic things found in an ecosystem:

- **Water.** Water is not alive, yet living things need water to survive. Water can be in the form of rain and snow, a river, a lake, the ocean, or water stored underneath the ground.

- **Rocks.** Rocks are an important part of ecosystems. Rocks can include the way the surface is shaped, with mountains and canyons. Tiny pieces of rocks are found in soil. These rocks have minerals that plants need in order to grow.

- **Air or gases.** Living things need the gases in air or water to survive. The gases are abiotic.

LIVING PARTS OF THE ECOSYSTEM

The living parts of an ecosystem are called **biotic**. Living things include all organisms found in the six different kingdoms of life: bacteria, archaebacteria, protists, fungi, plants, and animals.

Biotic parts of an ecosystem also include things that were once alive but are now dead and decaying. Dead and decaying organisms bring nutrients to soil, water, and other organisms that eat them.

Ecosystems have biotic and abiotic things. Can you name two biotic and two abiotic things in this ecosystem picture? The trees and the deer are biotic. The water, air, and rocks are abiotic.

HABITATS

A **habitat** is the specific place where an organism lives in its ecosystem. Organisms get the food, water, and shelter they need for survival from their habitat. Let's look at some different examples of habitats:

- A fox lives in a hole he digs in the ground. He hunts small animals in the surrounding forest.

- A squirrel makes a home in a tree. He builds a nest in a tree hole and stores food for the winter. He eats the tree's nuts, seeds, and shoots for food.

- A bird builds her nest in a tree. She flies through the air looking for insects and worms that she can swoop down and catch. She brings some of the food home for her babies.

- A snake lives in a hole in the ground. He slithers around and finds small mice, rats, or even a squirrel to eat.

- A bear lives in a cave. He walks to the river to snatch a salmon as it swims by. He will sleep in his cave at night, and it will protect him from the weather.

- Some fish will live in the ocean. It provides everything they need. They can find food to eat. Coral provides a hiding place from the bigger fish that want to eat them.

POPULATION

An organism does not live in an ecosystem all by itself. An anchovy fish does not swim the ocean alone. It swims with many other anchovies, called a school of fish. A dandelion will not be alone in a field. There will be many dandelions in a field. All the anchovies in a school are a population of anchovies. All the dandelions in a field are a population of dandelions.

A population of sardines

A **population** in an ecosystem is a group of one kind of organism living in one place at the same time. Population in ecosystems does not mean counting all the dandelions in a field. A population of dandelions in an ecosystem is all the dandelions in a field at the same time. Dandelions in one field are a different population than dandelions in someone's yard.

COMMUNITY

An ecosystem will have several different populations with different types of plants, animals, fungi, and bacteria. An ecosystem in a field will have populations of moles, grasses, wildflowers, dandelions, field mice, and mushrooms. This is a community. A **community** is all the different populations living in the same place at the same time.

A community in an ecosystem has many different populations.

Review

Match the words with the descriptions.

1.1) ____ organism

1.2) ____ interact

1.3) ____ habitat

1.4) ____ community

1.5) ____ abiotic

1.6) ____ ecosystem

1.7) ____ biotic

1.8) ____ population

A. all the different populations living in the same place at the same time

B. living

C. all the living and nonliving things that interact in a specific place

D. the place an organism lives in its ecosystem

E. nonliving

F. a group of one kind of organism living in one place at the same time

G. a living thing

H. to act in a way that affects something else

5

Write the correct answers.

1.9) What three things must an ecosystem have for organisms to be able to
live there?

    a. _____

    b. _____

    c. _____

    d. _____

1.10) What are three abiotic things found in an ecosystem?

    a. _____

    b. _____

    c. _____

1.11) What are the biotic things found in an ecosystem?

    a. _____ found in the six different kingdoms of life

    b. things that were once alive but are now _____ and

    _____

1.12) Organisms get the food, water, and shelter they need for survival from their

_____.

1.13) A(n) _____ of dandelions in an ecosystem would be all
the dandelions in a field at the same time.

1.14) An ecosystem in a field will have populations of moles, grasses, wildflowers,

dandelions, field mice, and mushrooms. This is a(n) _____.

1.15) The mudskipper in the picture lives in this puddle and finds its food, water, and shelter there.

This puddle is the mudskipper's

_____.

1.16) The picture shows trout fish living in a lake.

All the trout living in the lake are a(n)

_____.

1.17) There are many types of fish, plants, and other organisms living in the coral reef in the picture.

The coral reef is a(n)

_____.

Check ☐ Correct ☐ Recheck ☐

Write the correct answers.

1.18) How is a population different from a community? _____

_____

_____

_____

_____

_____

1.19) Describe the habitat of an animal you like or that you know a lot about. Be sure the habitat includes how the animal gets food, water, and shelter. Do research for more information if you need to. _____

_____

_____

_____

_____

_____

Teacher Check ☐

> I live in a fantastic biome—a forest. I love all the pine trees, the birds, and the sawdust gravy! The weather is perfect, not too hot and not too cold.

BIOMES

Ecosystems can be large or small. A **biome** is an ecosystem with certain plants, animals, and weather that covers a large area. Biomes have traits based on long-term weather patterns and the types of plants found there. The types of plants found in a biome will help determine the types of animals that live in a biome. Weather determines how hot or cold a biome will be and how much rain or snow a biome will get each year.

There are different types of biomes on earth. Every biome has plants and animals that are able to live there. Even though a desert looks very different from a jungle, both of these biomes have all the things that are necessary for life. Remember that there must be food, air, water, and shelter for an ecosystem to have plants and animals.

Let's briefly look at different biomes.

Desert

Deserts are dry. They get very little rain. Deserts get less than ten inches of rain per year. Most deserts are very hot. Many populations of plants and animals live in the desert, even though it is hot and dry.

Forest

Forest habitats cover much of earth's land surface. There are many different types of forest. Forests are found where there is a good amount of rainfall. Trees are very tall and need lots of water to live and grow.

One type of forest is called a temperate [TEM-per-it] forest. These types of forests are found in the United States. Temperatures are not too hot and not too cold. Many different types of trees and animals can be found in temperate forests.

Another type of forest is a tropical rain forest. Tropical rain forests have warm to hot weather and lots of rain. Tropical rain forests have more different types of plants and animals than any other ecosystem on land.

The taiga biome is found in colder places like Canada and Alaska.

The taiga [TAHY-guh] is a type of conifer forest. These forests are found in Canada and Alaska where the weather is cold for much of the year.

Grassland

Grasslands have less rainfall than forests and more rainfall than deserts. Temperate grasslands can be found in the United States. Some people call the grasslands the prairie. Corn and wheat grow well in the grassland areas of the United States.

Grasslands can also be found in Africa. The weather is hotter there than in the grasslands of the United States. Lions, elephants, and cheetahs all live in a grassland ecosystem. The grasslands found in Africa are called savannas.

Tundra

Tundra ecosystems are very cold with lots of ice. Plants are small and grow low to the ground. Animals are often white to blend in with the snow. Animals also have thick fur to help keep them warm.

Lakes and Rivers

Some ecosystems are found in water. Both lakes and rivers are filled with many populations of different organisms. There are algae. Algae are like plants and make their own food for energy. Reeds might be found on the edges of the lake or river. Reeds are able to have roots that are covered by water. Fish, frogs, dragonflies, beavers, and mosquitoes might be some animals in a lake or river ecosystem.

Ocean

Ocean ecosystems also have algae. There are sharks, fish, whales, dolphins, crabs, clams, coral, and starfish in ocean ecosystems. Animals that live in the ocean are able to live in salty water.

Reindeer live in the tundra biome.

Coral reefs are colorful ecosystems filled with many different types of organisms.

PLANT ADAPTATIONS

Plants in every biome need certain things to survive. They need light, usually in the form of the sun. Plants need minerals and water. Most plant roots take minerals and water from the soil. Plants need carbon dioxide *[dahy-OK-sahyd]* gas that is in the air.

Some plants have certain traits that help them survive in their biome. A trait that helps a plant or animal survive is called an **adaptation**. Let's look at some examples of some different traits that help plants survive in their ecosystem or biome.

Bromeliads

Bromeliads *[broh-MEE-lee-ads]* are found in many warm areas. Some bromeliads live in dry places. These plants can survive because of their special leaf adaptation. Their leaves form an open bowl-like shape that captures rain water and pools it for long periods of time. Even if it does not rain for a long time, the plant still has a source of water.

Lily pads live in water. Lily-pad leaves are large and flat. This helps the plant to be able to float on the water.

CLIMBING VINES

Plants that have vines are able to live in trees and other places where there is low light. They can do this because the vines will climb up the tree or building in order to reach the sunlight.

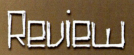

Write the correct answers.

2.1) What is a biome? Use a complete sentence. _____

_____

_____

_____

_____

2.2) The types of _____ found in a biome will help determine the

types of _____ that live in a biome.

2.3) _____ determines how hot or cold a biome will be and how

much rain or snow a biome will get each year.

2.4) _____ _____ _____ have

more different types of plants and animals than any other ecosystem on land.

2.5) What do plants need to survive?

    a. _____

    b. _____

    c. _____

    d. _____

2.6) What is an adaptation? Use a complete sentence. _____

_____

_____

_____

_____

2.7) The rose bush in the picture has thorns to help it survive by preventing animals from eating it.

This is an example of a(n)

_____.

Write D if the biome is a Desert, F for Forest, G for Grassland, T for Tundra, LR for Lake or River, or O for Ocean.

2.8) _____ has several types including temperate, tropical, and taiga

2.9) _____ might contain algae, reeds, fish, dragonflies, and beavers

2.10) _____ contains animals that are able to live in salty water

2.11) _____ might contain corn, wheat, lions, elephants, and cheetahs

2.12) _____ is very cold with a lot of ice

2.13) _____ contains plants that grow low to the ground and animals that are white and have thick fur

2.14) _____ is found where there is a good amount of rainfall

2.15) _____ is dry and hot with very little rain

2.16) _____ is sometimes called the prairie or savanna

3. ANIMAL SURVIVAL

Maile and Mia like to look at birds with their binoculars. Mia notices that birds have different shapes of beaks. She wonders why the beaks are shaped differently.

Animals, like plants, also have special adaptations to survive in their ecosystems. Animals need food, water, and shelter from their ecosystems in order to survive.

Animals can have body structures or behaviors that help them survive. A **structure** is a part of the body. Let's look at some of these adaptations.

BODY STRUCTURE

- The duck's webbed feet are perfect for paddling through water, but that is not all. The feathers are oily, which helps the duck float on water.

- A hawk has feet with talons on them. The talons help the hawk catch its food.

15

- A honeycreeper bird has a small, thin beak. The beak can easily get into small flowers and find nectar. A duck has a bill. A bill is good for eating fish, grass, insects, and many other different types of food.

- The spider makes a silk webbing. The spider uses it to build a web to trap insects. Some spiders even build webs to catch small birds or bats. The spider can also quickly make a line of silk it can use to drop out of its web if it feels threatened by another animal.

- Walruses live in areas where it is very cold. They have thick layers of fat called blubber. Blubber protects walruses from the cold. Walruses also have flippers that allow them to move on land and in water.

- The thorny devil lizard lives in the desert. This lizard has spikes all over its body that discourage an attacker. How does the thorny devil lizard get water in a dry desert? There are grooves along the lizard's back that form water from dew that collects in the desert at night. Dew is a type of liquid water that forms on the ground when water in the air is cooled into water droplets.

BEHAVIOR

Animals have many behaviors that help them survive. You already learned about some of these behaviors such as migration or hibernation to help survive cold winters.

One adaptation that some animals have is to live and work together.

Buffalo, zebras, horses, elephants, and antelope will live in herds. Living in a herd helps an individual in the population be able to survive another animal trying to eat it.

Wolves hunt in packs. By hunting together, they increase their chances of getting an animal to eat. Lions live and hunt together in a group called a pride. A pride of lions has one male lion, several female lions, and lion cubs. The male lion provides protection. The female lions work together to get their food and feed the young.

Dolphins and whales live together in groups called pods. A pod of dolphins can protect one another from an animal trying to eat one of them. A pod of dolphins can also get food and share the food with the young and baby dolphins.

Dolphins live together in a pod.

17

Some insects live in complicated groups. Ants live in a colony. Different types of ants have different jobs in the colony. The queen ant is responsible for laying eggs. Worker ants, which are female, help take care of the young and gather food. Male ants also help make baby ants. They don't have any other job to do.

These are just some adaptations that animals have to survive in their ecosystems. We will study more adaptations in a later Lesson.

Ants live in a colony.

Termite Colony

Termites, like ants, live in colonies. Some termites have a job to build the nest and take care of the young. Some termites are soldier termites. Their job is to protect the nest. Soldier termites have large jaws they use to protect the nest from enemies. The biggest termite colony in the world is found underground.

Fill in the blanks on the chart to describe the examples of animal adaptations.

Adaptation	How it helps the animal survive	Type of adaptation (body structure or behavior)
3.1) Ants live in a _____.	helps them take on different jobs that benefit the group	3.2) _____
3.3) A walrus has thick layers of fat called _____.	3.4) helps with protection from the _____	body structure
3.5) Ducks have _____ feet.	help with paddling through water	3.6) _____
3.7) Zebras live in _____.	helps them survive when animals try to eat them	3.8) _____
3.9) A hawk has feet with _____ on them.	3.10) help with catching _____	body structure
A spider makes silk webbing.	3.11) helps _____ insects	3.12) _____

19

3.13) Wolves hunt in _____	3.14) helps them increase the chances of getting an animal to _____	behavior
A thorny devil lizard has grooves on its back.	3.15) help collect water from _____ in the dry desert	3.16) _____.
3.17) Dolphins live together in _____.	3.18) helps them get _____ and share it with the young	behavior

Answer the question.

3.19) Think about an animal you have seen in the wild or learned about. What adaptations does the animal have to help it survive? Describe at least three adaptations. Do research for more information if you need to.

_____

_____

_____

_____

_____

_____

_____

Teacher Check

20

DUCKS

WEBBED FEET HELP DUCKS SWIM!

21

In this investigation, you will observe an ecosystem in your backyard or school yard. Be safe! Do not touch or pick any plants unless your teacher gives you permission. Make sure to wear closed-toe shoes. Wear gloves and safety goggles if you will be collecting any samples.

MATERIALS

- camera
- collecting net
- journal
- computer and printer to print out pictures

PROCEDURE

1 – In your journal, go to the next blank page. Title the page "Ecosystem Observation" and write the page number in the upper right. Go to the Table of Contents page and write the title and page number.

2 – Go outside in your backyard or schoolyard. Observe the ecosystem.

3 – In your journal, draw the different biotic and abiotic things you see.

4 – Take pictures with your camera. Take pictures of different plants and animals that you see.

5 – Use a collecting net to collect any insects you see. Carefully observe and take pictures of the insects, then set them free.

6 – Answer these questions in your journal:

   a. What type of biome are you observing?
   b. What is the community of this ecosystem? Are there dandelions, grasses, grasshoppers, flies, and wildflowers? Print out pictures and paste them into your journal to show the different organisms in the ecosystem.

GRADING CHECKLIST

- ❏ A drawing of the biotic and abiotic things in the ecosystem is in the journal.

- ❏ Pictures of the plants and animals in the ecosystem are pasted into the journal.

- ❏ The student collected, observed, and released insects from the ecosystem.

- ❏ The questions are answered in the journal.

 Teacher Check ❏

PICTURE THIS!

Flexibility is the willingness to change plans or ideas without getting upset. Helicopters are flexible because they fly in many directions. Unlike airplanes, helicopters can fly up, down, sideways, and even backward!

Just as a helicopter pilot adjusts to changing weather conditions, you can adapt to change and have a good attitude. Do not complain when things do not go your way. Look for the benefits of a change, and make the most of your situation.

Change will occur throughout your life. Learn flexibility today!

(Each answer, 4 points)
Fill in the blanks using words from the box below.

abiotic	adaptation	biome	biotic	community
ecosystem	habitat	interact	organism	population

1.01) A(n) _____ is a living thing.

1.02) To _____ is to act in a way that affects something else.

1.03) A(n) _____ is a place an organism lives in its ecosystem.

1.04) A(n) _____ is all the different populations living in the same place at the same time.

1.05) _____ means nonliving.

1.06) A(n) _____ is all the living and nonliving things that interact in a specific place.

1.07) _____ means living.

1.08) A(n) _____ is a group of one kind of organism living in one place at the same time.

1.09) A(n) _____ is an ecosystem with certain plants, animals, and weather that covers a large area.

1.010) A(n) _____ is a trait that helps a plant or animal survive.

Match the biomes with the descriptions.

1.011) ____ desert

1.012) ____ forest

1.013) ____ grassland

1.014) ____ lake or river

1.015) ____ ocean

1.016) ____ tundra

A. very cold with lots of ice

B. dry and hot with very little rain

C. sometimes called the prairie or savanna

D. populated by animals that are able to live in salty water

E. found where there is a good amount of rainfall

F. sometimes populated by algae, reeds, fish, dragonflies, and beavers

Choose the best answers.

1.017) _____ Which of these is an example of an animal adaptation of behavior?

A. Wolves hunt in packs.

B. Ducks have webbed feet.

C. A spider makes silk webbing.

D. all of these

1.018) _____ Which of these things found in an ecosystem is biotic?

A. rocks

B. organisms

C. water

D. all of these

1.019) _____ Which of these things found in an ecosystem is abiotic?

A. gases

B. rocks

C. water

D. all of these

1.020) _____ How does the blubber on the walrus in the picture help it to survive?

A. It makes attacks from other animals bounce off.
B. It helps the walrus to store up fat for hibernation.
C. It keeps the walrus warm in cold temperatures.
D. It helps the walrus to roll better so it can get away from other animals.

1.021) _____ Which of these is a reason why animals might want to live and work together?

A. They can do different jobs that benefit the group.
B. They can survive better when other animals try to eat them.
C. They can have a better chance of getting food.
D. all of these

1.022) _____ The picture shows an example of _____.

A. weather   C. a population
B. an adaptation   D. a community

1.023) _____ Which of these adaptations helps an animal to catch food?

    A. A hawk has feet with talons on them.

    B. A spider makes silk webbing.

    C. Wolves hunt in packs.

    D. all of these

1.024) _____ Which of these adaptations helps an animal to get water?

    A. A thorny devil lizard has grooves on its back.

    B. Dolphins live together in pods.

    C. Ducks have webbed feet.

    D. all of these

1.025) _____ All of the pine trees in the forest in the picture make up a(n) _____.

    A. biome

    B. adaptation

    C. population

    D. community

Check ☐ Correct ☐ Recheck ☐

OBJECTIVES:

- Define producer and consumer.
- Draw a diagram of photosynthesis.
- Understand that plants need water, sunlight, and carbon dioxide to do photosynthesis.
- Understand how energy flows in a food chain and a food web.
- Understand the difference between a food chain and a food web.
- Describe predator and prey.
- Describe different adaptations that predators and prey have that help them survive.

VOCABULARY:

camouflage *[KAM-uh-flahzh]* – (noun) the ability to blend into the background of the environment

chlorophyll *[KLAWR-uh-fil]* – (noun) a green material used by plants to take in the sun's energy and turn it into sugar

consumer *[kuhn-SOO-mer]* – (noun) an organism that eats another organism for food

mimicry *[MIM-ik-ree]* – (noun) the ability of an animal or plant to look like a more dangerous predator

photosynthesis *[foh-tuh-SIN-thuh-sis]* – (noun) a special process that plants use to make their own food

predator *[PRED-uh-ter]* – (noun) an animal that eats another animal

prey – (noun) an animal that gets eaten

producer *[pruh-DOO-ser]* – (noun) an organism that makes its own food

4. PRODUCERS AND CONSUMERS

Macy loves to eat popcorn. She wishes she could only eat popcorn. Her mom makes her eat fruits, vegetables, and meat though.

PRODUCERS

In every ecosystem, each organism needs energy to live. Energy to live comes in the form of food. Organisms are classified as producers or consumers based on how they get food.

Producers, like these daisies, make their own food.

A **producer** is an organism that can make its own energy, or food. Plants, some bacteria, and some protists are producers.

Plants make their own food in a special process called **photosynthesis**. Photo means light and synthesis means to make. Photosynthesis takes carbon dioxide, sunlight, and water and makes them into a sugar.

Carbon dioxide is a gas found in the air. One way carbon dioxide gets in the air is when animals and plants breathe out. Every time you breathe out, you breathe carbon dioxide into the air. Carbon dioxide also gets into the air when cars and trucks burn gas. Cars have pipes that release gases into the air. Carbon dioxide comes out of these pipes.

Plants then take the carbon dioxide inside their leaves. The leaves also take in water that is in the air. Sunlight or another type of light also must hit the leaves. Photosynthesis does not happen at night.

Plants take the energy in sunlight, water, and carbon dioxide and make two things: sugar and oxygen gas.

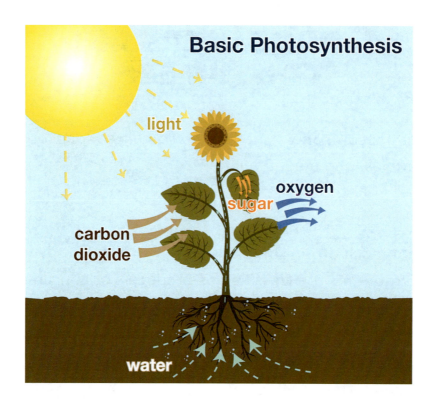

Basic Photosynthesis

Animals and people need both the food and oxygen that plants provide. When we breathe air in, our bodies use oxygen gas. People also eat the sugars that plants make. Sometimes the sugars are stored in the stems and fruits of plants. Sometimes the plant turns sugar into something else we eat.

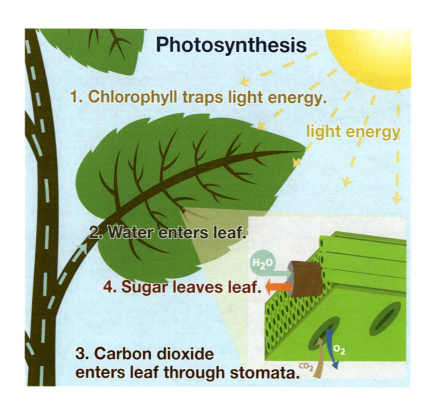

Photosynthesis

1. Chlorophyll traps light energy.

light energy

2. Water enters leaf.

4. Sugar leaves leaf.

3. Carbon dioxide enters leaf through stomata.

Sometimes we eat animals that eat plants. We could not eat a hamburger that comes from cows without plants because cows eat grass.

Let's look more closely at how a leaf makes sugar.

Pipes that travel throughout the plant carry water and minerals from the roots into the leaves. These tiny pipes are called veins. They are similar to the large and small veins that carry blood in our bodies. Have you ever wondered why most plants are green? This is because of something called **chlorophyll**. Chlorophyll is green. It is used by plants to take in the sun's energy and turn it into sugar.

There are many holes on the underside of a leaf. These holes are called stomata [STOH-muh-tuh]. The stomata take in carbon dioxide gas. They also allow oxygen gas to go back into the air during photosynthesis.

CONSUMERS

Consumers must eat another organism for energy, or food. You have already learned about some types of consumers: herbivores, carnivores, omnivores, and decomposers.

Remember, some organisms can be more than one type of consumer.

Scavengers

A scavenger is a special type of decomposer. It is an animal that eats dead animals. A scavenger cannot be a plant or fungus. Scavengers do not feed on dead plants, like fungi or bacteria do.

Hyenas, vultures, crows, and hagfish are all examples of scavengers. Scavengers will eat a dead animal that died naturally, or

they will eat what is left over after another animal kills and eats. Hyenas will wait until lions have killed a zebra or an antelope. When the lion is done eating, the hyenas will then eat what is left over.

Write the correct answers.

4.1) What is a producer? Use a complete sentence. _____

_____

_____

4.2) List three examples of organisms that are producers or are sometimes producers.

a. _____

b. _____

c. _____

4.3) What is photosynthesis? Use a complete sentence. _____

_____

_____

4.4) Photosynthesis takes _____ _____,

_____, and _____ and makes them into

a sugar.

4.5) When does carbon dioxide gas get into the air?

a. when animals and plants _____

b. when cars and trucks _____ _____

4.6) What do plants make during photosynthesis?

a. _____

b. _____

4.7) Label the diagram of photosynthesis.

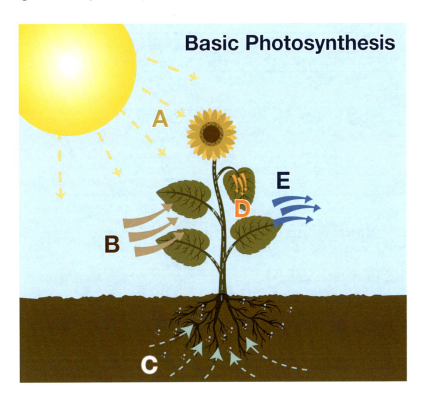

A – _____

B – _____

C – _____

D – _____

E – _____

4.8) Plants are green because of _____.

4.9) What is chlorophyll? Use a complete sentence. _____

_____

_____

_____

4.10) The holes on the underside of a leaf are called _____.

4.11) What do stomata do?

 a. take in _____ _____ gas

 b. allow _____ gas to go back into the air

4.12) What is a consumer? Use a complete sentence. _____

_____

_____

4.13) What are four different types of consumers?

 a. _____

 b. _____

 c. _____

 d. _____

Answer the question.

4.14) What is the difference between scavengers and other types of decomposers?

_____

_____

_____

5. FOOD WEBS

"Ahh...the Sun. It's wonderful. I get energy from the sun—sort of. The flowers get energy from the sun and I get energy from flowers! So without the sun, I'd have no energy to do any work."

Organisms need energy to live and grow. Producers make their own energy, and consumers eat something else for energy. In an ecosystem, you can study how this energy flows or moves from one organism to another organism.

FOOD CHAIN

One way to show how energy flows in an ecosystem is a food chain.

A food chain shows a direct line of energy. A food chain starts with a producer. The producer takes the energy from the sun and converts it into sugar and other types of food.

An animal (consumer) will eat the producer (usually a plant) and get energy from it. Another animal could then eat that animal and get energy from it. Let's look at a food chain you might find in a desert ecosystem.

Desert Food Chain

| plants | ants | lizard | snake | hawk |

An ant will eat a desert plant. A lizard will eat the ant. A snake eats the lizard. A hawk eats the snake.

FOOD WEB

Food chains are very simple. They do not show all the organisms a lizard eats. A food web is a better way to show what eats what in an ecosystem. A lizard does not just eat ants. A lizard will eat termites, crickets, flies, and many other kinds of insects. A hawk will eat mice, snakes, small rabbits, and fish.

A food web will show energy flow between many different plants and animals. There are arrows going out from the animal or plant to what eats it. The arrows show which way the energy is moving.

Look at this food web.

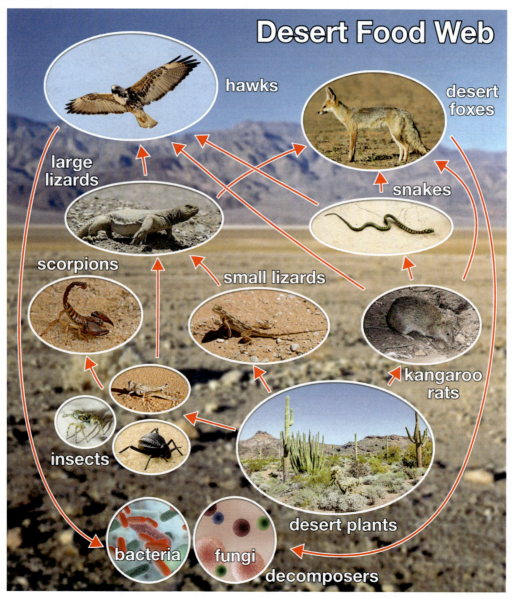

Desert Food Web

hawks

desert foxes

large lizards

snakes

scorpions

small lizards

kangaroo rats

insects

desert plants

bacteria   fungi

decomposers

We can see that a desert fox eats large lizards, snakes, and kangaroo rats. We can see that desert plants are eaten by kangaroo rats, small lizards, and insects.

Notice that decomposers are part of the food web. Decomposers are usually bacteria and fungi. Decomposers are important. When decomposers eat dead and decaying matter, they put nutrients back into the soil, water, or sand. Plants need these nutrients to be able to grow.

Let's look at an ocean food web. Even in water, there needs to be producers and decomposers. Most producers in the ocean are algae. Algae are plant-like protists. One type of algae is a bacteria. Algae are able to do photosynthesis like plants. Algae, like plants, need energy from the sun.

Ocean Food Web

blue whale • killer whale • Ross seal • crabeater seal • antarctic petrel • leopard seal • emperor penguin • Weddell seal • herring • krill • Adélie penguin • squid • anchovies • algae

○ carnivore (consumer)
○ herbivore (consumer)
○ producer

HOW COMPLEX IS YOUR FOOD WEB?

People are also part of food webs. All around the world, people eat many different types of plants and animals. Some communities eat foods that many other cultures do not. For example, in parts of Asia, some people eat dog meat. In France, people eat snails. In Louisiana, some people eat alligator. These meats are considered exotic because they are not found in the meat section of your supermarket next to hamburger meat, bacon, and hot dogs. Think about the different types of plants and animals you eat. Do you eat many different kinds? Or do you like to eat only a few items, like apples and chicken nuggets?

Write the correct answers.

5.1) A(n) _____ _____ shows a direct line of energy.

5.2) Fill in each blank on this food chain diagram with the word producer or consumer.

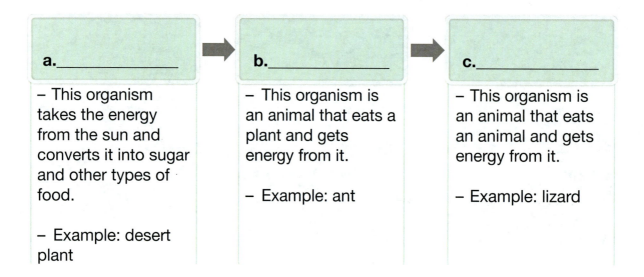

a._____
– This organism takes the energy from the sun and converts it into sugar and other types of food.

– Example: desert plant

b._____
– This organism is an animal that eats a plant and gets energy from it.

– Example: ant

c._____
– This organism is an animal that eats an animal and gets energy from it.

– Example: lizard

5.3) On the diagram, if more organisms are added to the right side of the food chain, will they be producers or consumers? _____

5.4) A(n) _____ _____ will show energy flow between many different plants and animals.

5.5) When _____ eat dead and decaying matter, they put nutrients back into the soil, water, or sand.

5.6) Most producers in the ocean are _____, which are protists or bacteria.

Check ☐ Correct ☐ Recheck ☐

Write the correct answers.

5.7) Why is it better to use a food web instead of a food chain?

_____

_____

_____

_____

5.8) Turn to the next blank page in your journal and title it "Food Web." Put the page number in the upper right. Add the title and page number to the Table of Contents page. Use the chart to draw a food web showing how energy flows between the organisms in this ecosystem. There should be an arrow pointing from each organism to the organism that eats or uses it. When you are finished, add a human to the food web. Write producer, consumer, or decomposer next to each organism in the food web.

Organism	What the Organism Eats or Uses
grass	uses nutrients from worms
grasshopper	eats grass
cow	eats grass
rabbit	eats grass
bird	eats grasshoppers and worms
coyote	eats cows, rabbits, and birds
worm	breaks down dead grass, grasshoppers, cows, rabbits, birds, and coyotes

Teacher Check ☐

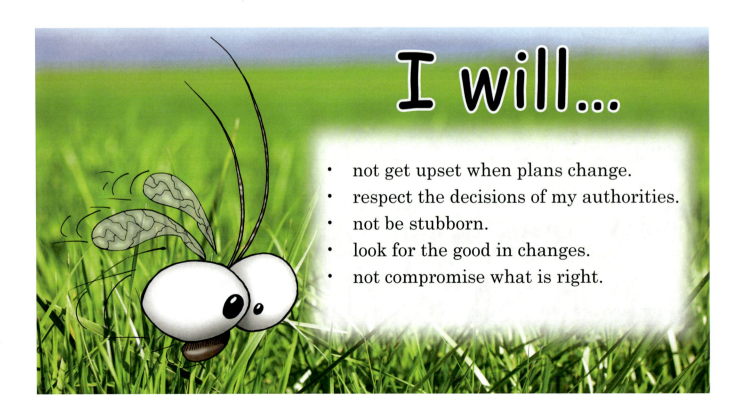

I will...

- not get upset when plans change.
- respect the decisions of my authorities.
- not be stubborn.
- look for the good in changes.
- not compromise what is right.

6. PREDATOR AND PREY

Lakisha, her dad, and her dog named Buddy are walking in the park. The dog starts barking at a noise in the trees. The dog gets loose from the leash and runs toward the trees. He comes back whining and stinking! Lakisha's eyes start to water. "Buddy has been sprayed by a skunk!" her dad says. "It's going to take several days to get the stink off of Buddy. I hope he learned his lesson and won't chase skunks anymore."

You have learned that plants and animals have many different names within an ecosystem. There are producers and consumers. There are herbivores and carnivores. In this Lesson, you will learn about two more names for organisms in ecosystems: predator and prey.

When you look at a food web or food chain, there are organisms that get eaten and organisms that do the eating.

An animal that gets eaten is called the **prey**. An animal that eats another animal is called a **predator**. Sometimes an animal can be both a prey and a predator.

Let's look at this food web for a tropical rain forest.

The golden lion tamarin eats leafcutter ants. The golden lion tamarin is the predator and the ants are the prey. The ocelot *[OS-uh-lot]* eats the golden lion tamarin. Now the golden lion tamarin is a prey and the ocelot is a predator.

sun

anaconda
snake

ocelot

poison
dart frog

golden lion
tamarin

**The Rainforest
Food Web**

leafcutter
ants

leaves

fruit tree

PREDATOR AND PREY ADAPTATIONS

Predator Adaptations

Some animals have adaptations that make them good predators. Let's look at some examples of predator adaptations:

- **Speed and strength.** Cheetahs run very fast. Their speed makes them good predators. They can run faster than any other animal. Lions are powerful and strong. They are able to take down large prey.

- **Camouflage. Camouflage** is the ability to blend into the background of the environment. Camouflage helps a predator sneak up to its prey.

- **Poison.** Poison is a type of matter that causes harm. Poison is usually a liquid. The poison dart frog is able to poison its prey, which causes them to not be able to move. Some snakes have a poisonous venom. The venom goes from the snake's fangs into its prey. The poison will either keep the prey from moving or kill the prey.

Prey Adaptations

Many organisms also have adaptations that help them not get eaten.

- **Spikes and thorns.** A hedgehog is covered with spikes. When a hedgehog feels threatened by a predator, it curls into a ball. When it curls into a ball, only the spikes show. This helps protect the hedgehog, because predators do not want to be poked by the spikes.

Roses, cacti, and many types of plants have thorns. Plants are not prey. However, thorns make animals not want to eat the plant.

- **Smell.** Have you ever smelled a really strong, horrible smell? Your mom or dad might say, "That's a skunk!" Skunks give off a strong odor when scared. This smell keeps predators away. The strong smell says to the predator, "Don't eat me! I don't taste good."

- **Camouflage.** Some types of prey also have camouflage. If it is harder to see an animal, it is harder for the animal to become prey.

- **Mimicry. Mimicry** is the ability of an animal or plant to look like a more dangerous predator. The animal will not be harmful, but other animals will think it is dangerous and keep away. One example is the coral snake and the king snake. The coral snake has poisonous venom. The king snake does not have poisonous venom. The king snake looks like the coral snake. Predators might not want to eat a king snake because they would think it was the poisonous coral snake.

Coral snake

Plants and animals want to survive, predators want to be able to get food, and prey want to avoid being eaten.

King snake

44

Use the food chain to answer the questions.

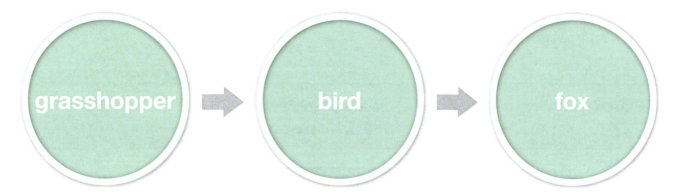

6.1) Which organisms are prey?

    a. _____

    b. _____

6.2) Which organisms are predators?

    a. _____

    b. _____

6.3) Which organism is both a predator and a prey? _____

Fill in the blanks.

6.4) A(n) _____ is an animal that gets eaten.

6.5) A(n) _____ is an animal that eats another animal.

6.6) Sometimes an animal can be both a(n) _____ and a(n) _____.

6.7) _____ is the ability to blend into the background of the environment.

6.8) _____ is the ability of an animal or plant to look like a more dangerous predator.

Fill in the blanks on the charts.

Predator adaptation	How it helps the predator get food	Examples
6.9) _____ and _____	The predator can run fast to catch food or attack large food.	6.10) _____; lion
camouflage	6.11) The predator can _____ up to its prey.	tiger
6.12) _____	The predator can keep the prey from moving or kill the prey.	6.13) poison dart _____; snake

Prey adaptation	How it helps the prey not get eaten	Examples
6.14) _____ and _____	Predators do not eat the prey because they do not want to be poked.	6.15) _____; rose; cactus
smell	6.16) This makes the predator think the prey does not _____ good.	6.17) _____

6.18) _____	This makes it harder for the predator to see and find the prey.	leaf insect
6.19) _____	6.20) Predators will think the prey is _____ and keep away.	king snake

Check ▢ Correct ▢ Recheck ▢

I will look for good in changes,
And will not fuss or fret.
I will trust in my authorities
And not get all upset.
I will bend where it is needed
By not hanging on too tight.
But will not be made to compromise
The things I know are right.

ACTIVITY
DESIGN AN AQUARIUM ECOSYSTEM

In this investigation, you will design an aquarium ecosystem. You will then observe the aquarium ecosystem for two weeks.

MATERIALS

These materials are all suggestions. The final setup will depend on your design.

- aquarium set up (This should include filter, pump, and heater.)
- rocks, gravel, and sand
- different types of fish, plants, and other organisms
- decorations
- fish food

PROCEDURE

Part 1

1 – In your journal, go to the next blank page. Title the page "Aquarium Design" and write the page number in the upper right. Go to the Table of Contents page and add the Activity title and page number.

2 – Think about everything that needs to go into your aquarium for the plants and animals to survive in that ecosystem. Write these things down in your journal.

3 – Write down what types of organisms will be in your aquarium.

4 – Draw a picture of your aquarium with all the materials you have decided to put in it.

5 – Get teacher approval on your design before beginning Part 2.

Part 2

6 – Set up your aquarium with your design.

7 – Observe the ecosystem for two weeks.

8 – Write your observations from the two weeks in your journal.

9 – At the end of two weeks, write a report. The report should include these things:

- a drawing of your aquarium

- a list of the parts of the aquarium that helped the organisms survive

- a list of organisms that were producers and a list of organisms that were consumers

- a drawing of a food web that represents your aquarium

- any observations you thought were important

- any ideas about how you can improve the design of your aquarium

10 – Share your report and findings with your teacher.

GRADING CHECKLIST

❑ The aquarium design was approved by the teacher.

❑ The aquarium was set up based on the design.

❑ The student observed the ecosystem for two weeks and recorded observations.

❑ The report was completed thoughtfully.

Teacher Check ☐

Each answer, 4 points)
Match the names with the parts of photosynthesis.

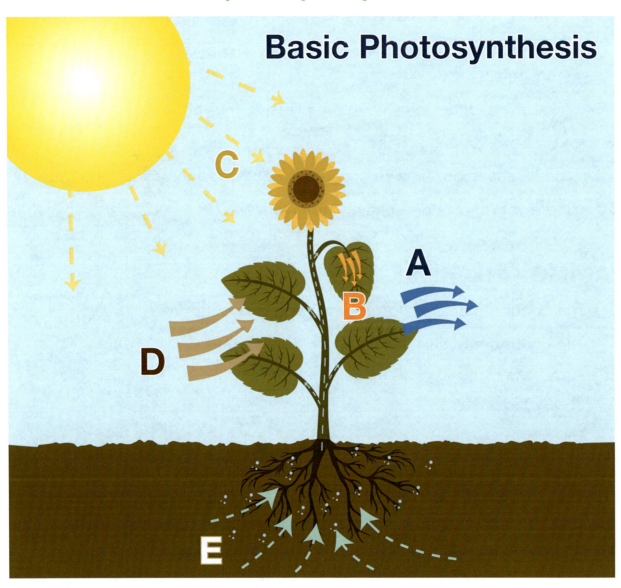

Basic Photosynthesis

2.01) ____ carbon dioxide

2.02) ____ light

2.03) ____ oxygen

2.04) ____ sugar

2.05) ____ water

Use the diagram to choose the best answers.

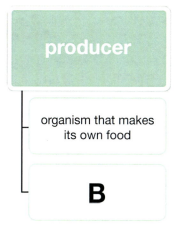

producer — organism that makes its own food

B

A — organism that eats another organism for food

herbivores, carnivores, omnivores, and decomposers

2.06) _____ What belongs in box A?

    A. photosynthesis

    B. scavenger

    C. consumer

    D. prey

2.07) _____ What belongs in box B?

    A. plants, some bacteria, and some protists

    B. plants, animals, and some bacteria

    C. protists and bacteria

    D. plants, some animals, and some bacteria

Use the food web to choose the best answers.

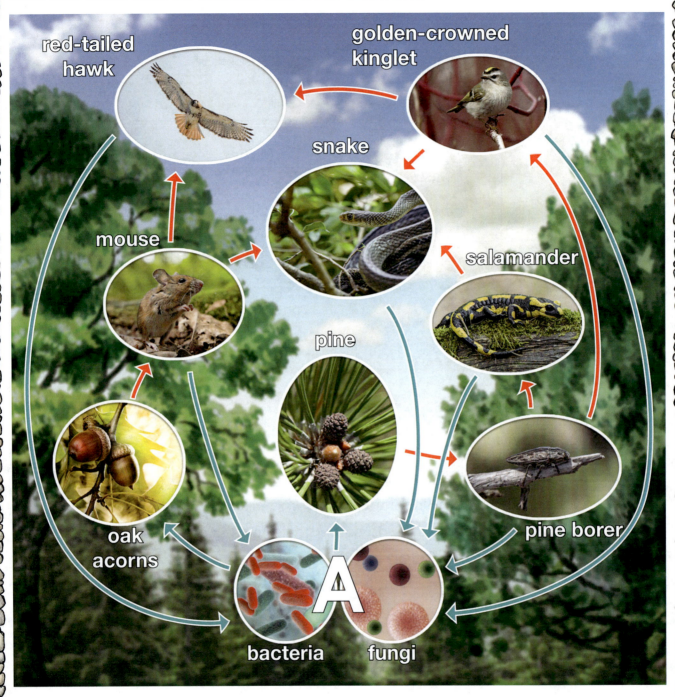

2.08) _____ What type of organisms are in the circles labeled A?

    A. producer   B. consumer   C. decomposer   D. herbivore

2.09) _____ What type of organism is the red-tailed hawk?

    A. producer   B. consumer   C. decomposer   D. herbivore

2.010) _____ What type of organism is the pine?

                A. producer   B. consumer   C. decomposer   D. herbivore

Match the words with the descriptions.

2.011) _____ camouflage

2.012) _____ chlorophyll

2.013) _____ food chain

2.014) _____ food web

2.015) _____ mimicry

2.016) _____ photosynthesis

2.017) _____ predator

2.018) _____ prey

2.019) _____ stomata

A. a green material used by plants to take in the sun's energy and turn it into sugar

B. a special process that plants use to make their own food

C. the ability to blend into the background of the environment

D. way to show a direct line of energy flow in an ecosystem

E. an animal that gets eaten

F. an animal that eats another animal

G. way to show energy flow between many different plants and animals in an ecosystem

H. the ability of an animal or plant to look like a more dangerous predator

I. holes on the underside of a leaf that take in carbon dioxide gas and allow oxygen gas to go back into the air

Choose the best answers.

2.020) _____ The penguin in this food chain are _____.

         A. consumers   B. predators   C. prey   D. A and B

2.021) _____ How does camouflage help a predator to get food?

A. It helps the predator to sneak up to the prey.

B. It makes it harder for the predator to find the prey.

C. It helps the predator to look more dangerous than it is.

D. It allows the predator to keep the prey from moving.

2.022) _____ What do plants make during photosynthesis?

A. carbon dioxide gas and water

B. carbon dioxide gas and sugar

C. sugar and oxygen gas

D. sugar and light

2.023) _____ What do plants take in during photosynthesis?

A. carbon dioxide gas, light, and water

B. sugar, light, and water

C. oxygen gas, carbon dioxide gas, and sugar

D. oxygen gas, water, and light

2.024) _____ Which of these is an adaptation that helps a predator get food?

A. poison   C. mimicry

B. smell   D. all of these

2.025) _____ Which of these is an adaptation that helps a prey not get eaten?

A. camouflage   C. smell

B. spikes and thorns   D. all of these

Check ☐   Correct ☐   Recheck ☐

OBJECTIVES:

- Describe different ways ecosystems can change.
- Describe how ecosystem change can cause some organisms to be endangered or extinct.
- Understand how fossils help scientists learn how change affects some organisms.
- Describe different ways to conserve natural ecosystems.

VOCABULARY:

chemical *[KEM-i-kuhl]* – (noun) a type of matter that can interact with other types of matter

conservation *[kon-ser-vey-shuhn]* – (noun) the act of protecting or restoring a natural environment or ecosystem

conserve *[kuhn-SURV]* – (verb) to protect from harm

emit *[ih-MIT]* – (verb) to put out

endangered *[en-DEYN-jerd]* – (adjective) describes an organism whose population is in danger of dying out

extinct *[ik-STINGKT]* – (adjective) describes an organism that used to live on earth but whose population completely died out

fossil *[FOS-uhl]* – (noun) the remains of a plant or animal

imprint fossil – (noun) a fossil that is the mark of a plant or animal

model *[MOD-l]* – (noun) an object that shows how something works or what something looks like

7. ECOSYSTEM CHANGE

Billy the goat doesn't like change. He likes everything to stay the same. He likes to run along the same paths eating from the same bushes and trees every day. One day, his family goes to a new place. There is a different type of tree there. It's new. It's different. He decides to climb the tree and eat the leaves. "Hmmmmm. This is delicious," he says. "Sometimes change is good!"

Ecosystems are always changing. Plants and animals come to a new place. Storms and fires can destroy ecosystems. People come to new places and build buildings, destroying habitats. Pollution in the soil, water, and air can harm or kill plants and animals.

There are two types of ecosystem change—natural and man-made.

NATURAL CHANGES

Let's look at some natural changes that can occur in an ecosystem:

- **Fire.** Fire can change an ecosystem. While some fires are caused by people, many fires are natural. Lightning is one way a natural wildfire starts.

Fire may seem very bad for an ecosystem. However, many ecosystems depend on natural wildfires. Fires clear away small bushes and trees. Some types of trees have seeds that will only germinate once there has been a fire.

Forest fires can help some ecosystems.

- **Storms.** Storms such as hurricanes can cause flooding. If plants are underwater for a long period of time, they might die. Animals can also die in flooding. Sometimes flooding carries animals to a new location. The animals in the new location might change the ecosystem they will now call their home.

Drought harms ecosystems, causing plants to die and animals to leave.

- **Drought.** Drought happens when an ecosystem gets less rainfall than normal for a long period of time. Drought can change ecosystems. Animals might move to another location to look for water. Many plants will not be able to survive without enough water. In some places on earth, droughts are causing some grassland ecosystems to turn into deserts.

MAN-MADE CHANGES

Some ecosystem changes are caused by people.

Building. People need places to live. Towns and cities where people live were once forests and grasslands. People changed the ecosystem to meet their needs. A shopping mall was once a habitat for plants and animals. When the shopping mall was built, these animals had to find another place to live or they would not survive.

Using natural resources. People use natural resources like metals, rocks, and minerals. Companies take these resources from the ground. The surface of the earth can be destroyed, changing the ecosystem that was once there.

When cities are built, natural ecosystems are changed.

Trees are another natural resource. Trees provide wood for building and material for paper. When we cut down forests, we take away the habitat for many different organisms.

Catching a ride. People move and travel all over the earth. Some plants and animals catch a ride with people. Sometimes people take plants and animals to a new place on purpose. Many times people bring plants and animals to a new place by accident.

When an animal is in its natural ecosystem, there is balance. In its natural ecosystem, there will be a predator that will keep the animal's population numbers low. When they reach a new place, some animals have no predators. These animals can reach very high population levels. They might eat all the plants and leave no food for other animals.

Many scientists work hard at keeping these plants and animals from taking over and harming new ecosystems.

The nutria is an animal that is taking over new ecosystems. The nutria's natural ecosystem is in South America. The nutria came to the United States in 1889 because some farmers wanted to raise them for their fur. Some nutria escaped into the wild. They do not have any natural predators in North America and quickly took over ecosystems. There are about 20 to 30 million nutria in Louisiana alone! Nutria have been found in 22 states. Scientists are working hard to keep nutria out of ecosystems in the United States.

Pollution. Humans can pollute ecosystems by putting harmful things in the air, water, or soil. When an ecosystem is polluted, it can quickly change. Here are some examples of how people pollute ecosystems:

- People can throw their trash on the ground. Birds and other animals can eat this trash. The trash can make animals sick and even kill them.

The litter on this beach is harmful to the beach ecosystem.

- Ships carrying oil can leak or spill the oil into the ocean. The oil gets stuck on birds' feathers. Clams and oysters get oil inside them. Some animals die because they have too much oil on them. The oil can make other animals and people sick if they eat animals that have been covered in oil.

This turtle is covered in oil.

Smoke from cars can pollute the air.

- People can put harmful liquids and solids into the ground or water. The plants can die. Animals might get sick if they drink the water. Animals could also get sick if they eat plants that are growing in a place with harmful liquids and solids. One harmful type of matter is pesticides. Pesticides are used when growing crops to kill the insects that might eat the crops. Too much pesticides can be harmful to an ecosystem.

- Factories and cars put smoke in the air. The smoke can make people and animals sick. Sometimes the smoke becomes part of the rain. The rain can then fall to the ground, polluting it.

Write the correct answers.

7.1) What are the two types of ecosystem change?

a. _____

b. _____

7.2) List three natural ecosystem changes.

a. _____

b. _____

c. _____

7.3) List three man-made ecosystem changes.

a. _____

b. _____

c. _____

7.4) What are pesticides? _____

_____

_____

_____

_____

Match the ecosystem changes with the descriptions.

7.5) _____ building

7.6) _____ catching a ride

7.7) _____ drought

7.8) _____ fire

7.9) _____ pollution

7.10) _____ storm

7.11) _____ using natural resources

A. Plants and animals lose their habitats when people use land for towns and cities.

B. People use metals, rocks, minerals, and trees, which can destroy the surface of the earth and take away habitats.

C. People put harmful things in the air, water, or soil, which can cause plants to die and animals to get sick.

D. When an area gets less rainfall than normal for a long period of time, plants might die and animals might move to another location to look for water.

E. Some types of trees have seeds that will only germinate when this change happens.

F. This type of change can cause flooding, which might make plants die and animals move to a new location.

G. People take plants and animals to a new place on purpose or by accident, and the animals might reach very high population levels.

Fill in the blanks on the chart to show how pollution hurts ecosystems.

Example of Pollution	How It Hurts the Ecosystem
7.12) People can throw their _____ on the ground.	7.13) The trash can make animals _____ and even kill them.
7.14) Ships carrying oil can leak or spill the oil into the _____.	7.15) Some animals die because they have too much _____ on them. Animals or people can get sick if they _____ the animals covered in oil.
7.16) People can put harmful _____ and _____ into the ground or water.	7.17) Plants can die and animals can get sick if they _____ the water or _____ the plants growing where the harmful things are.
7.18) Factories and cars put _____ in the air.	7.19) The smoke can pollute the air and make people and animals _____. If the smoke becomes part of the rain, it can fall to the ground, _____ it.

Check ☐   Correct ☐   Recheck ☐

Answer the question.

7.20) Why would it be bad if animals were taken to a new place where they had no

predators? _____

_____

_____

_____

_____

_____

_____

Teacher Check

WINDS OF CHANGE

The wind will tear a pinwheel that does not turn. Similarly, a person who does not adjust to change will be frustrated. Make your own pinwheel as a reminder to be flexible.

Supplies:
- heavy paper
- plastic drinking straw
- straight pin with a head
- thin-nosed pliers
- scissors
- pen or pencil

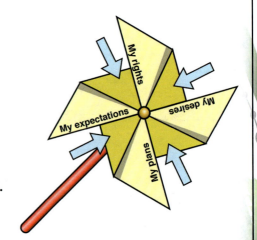

Cut a sheet of heavy paper into a square. Cut half the distance to the center from each corner.

On the corner of the triangles, write "My Expectations," "My Rights," "My Plans," and "My Desires."

Bend the other four corners into the middle of the paper and hold them there with a straight pin. Poke the pin through the straw and bend down with pliers to lie flat. Tape the point to the straw so that it will not poke anyone. Adjust the pinwheel so that it spins freely.

Points to Ponder:

- Flexibility is changing plans without getting upset. You can plan to see your friends this afternoon, but something might happen to your family that requires you to change your plans.

- A flexible person is willing to change, but not if it goes against the standards of good character. You can play with your friends, but not if it involves stealing or breaking something that does not belong to you.

8. ENDANGERED PLANTS AND ANIMALS

Betty is digging a hole in her yard. She is sure if she could dig a deep enough hole, she would find dinosaur bones! She imagines what it would be like to see dinosaurs running down her street.

Some plants and animals cannot survive changes in ecosystems. **Endangered** means that the population of a plant or animal is about to become extinct or is in danger of dying out. When an animal or plant is **extinct**, it means that no living organisms of that plant or animal remain on earth.

Pollution and disappearing habitats can all cause plants and animals to become endangered. Some animal populations become low from overhunting by humans. Other plant and animal populations cannot survive a new predator that comes into their habitat.

In the United States, some plants and animals are on an Endangered Species List. The Endangered Species List shows plants and animals whose population numbers are low. These plants and animals will have their habitats protected, and money is spent trying to keep them from becoming extinct.

The California condor is on the Endangered Species List. This bird is the largest bird in North America. The California condor is a scavenger. It likes to feed on large, dead animals. Half of the condors in the wild die of lead poisoning. They get lead poisoning by eating bullets made from lead that people use to shoot animals. The bullets remain in the dead animals and the California condor ends up eating the bullets along with the animal remains.

In 1987, the last condors (less than 20) in the wild were taken into captivity. In

The Lake Erie watersnake was once on the Endangered Species List. Population numbers have risen so much that the snake was taken off the list.

captivity, they reproduced. People also made sure they had enough food. Now, there are over 400 condors in the world. Half of the condors are in captivity, like in a zoo. The other half live in the wild. If the California condor goes extinct, things can get messy. As a scavenger, they are nature's cleanup crew. This scavenger plays a very important part in its ecosystem.

FOSSILS

Scientists study plants and animals that have already gone extinct. They hope that by studying the past, they can help prevent extinction of plants and animals today.

One way to study extinct plants and animals is to study fossils. **Fossils** are the remains of plants and animals.

Most fossils are very old. Dinosaur fossils tell scientists about these giant reptiles that lived on earth long ago.

Scientists studied this triceratops fossil. From studying the fossil, they can predict how much it weighed and what it ate. They can make a model of what it looked like.

A **model** shows how something works or what something looks like. A triceratops model would represent what a triceratops might have looked like when it was alive. A solar system model would be a smaller representation of the sun and the planets. Many models are smaller than the object or system they are showing.

Models made from fossils have limitations *[lim-i-TEY-shuhns]*. A limitation is something that keeps the fossil from being exactly like the thing it represents. One limitation of models from fossils is that many times there is not a complete fossil skeleton of an animal. To complete the models, people have to figure out what the other parts of the skeleton might look like.

A fern has been imprinted in this fossil. While no parts of the fern remain, you can see still see what the fern looked like.

Most fossils are the remains of the hard parts of animals, like bones and teeth. Sometimes an entire animal can be found as a fossil. Sometimes only part of the animal, like a skull, remains.

Other times, only a mark the plant or animal made remains. A mark can be a footprint or the pattern of a leaf. A fossil that is the mark of a plant and animal is called an **imprint fossil**.

Most fossils are found inside rocks and soil. Fossils found in rock form when a plant or animal dies in mud. Over many years, the mud turns into rock. Sometimes the bones and teeth become fossils in the mud. Other times, the mark a plant or animal left in the mud becomes the fossil.

Some fossils are found in a plant material called amber. Other fossils are found inside ice.

This spider was fossilized in amber.

Review

Match the words with the descriptions.

8.1) _____ endangered

8.2) _____ Endangered Species List

8.3) _____ extinct

8.4) _____ fossil

8.5) _____ imprint fossil

8.6) _____ model

A. describes an organism that used to live on earth but whose population completely died out

B. an object that shows how something works or what something looks like

C. the remains of a plant or animal

D. describes an organism whose population is in danger of dying out

E. something used in the United States that shows plants and animals whose population numbers are low

F. a fossil that is the mark of a plant or animal

Write the correct answers.

8.7) List three things that can cause plants and animals to become endangered.

a. _____

b. _____

c. _____

8.8) Animals on the Endangered Species List will have their habitats

_____, and _____ is spent trying

to keep them from becoming extinct.

8.9) Why do scientists study plants and animals that have already gone extinct?

_____

_____

_____

8.10) Most fossils are the remains of the hard parts of animals, like

_____ and _____.

8.11) What are two examples of marks that can turn into imprint fossils?

a. _____

b. _____

8.12) What are four materials that fossils are found inside?

a. _____

b. _____

c. _____

d. _____

8.13) Fossils found in rock form when a plant or animal dies in _____, and over many years, the mud turns into _____.

 Check ☐ Correct ☐ Recheck ☐

Write the correct answer.

8.14) Explain how being on the Endangered Species List helped the California condor.

_____

_____

_____

_____

_____

_____

_____

 Teacher Check ☐

9. CONSERVATION

Imagine if the whole earth was covered by one big city. Everywhere you looked, there were streets and buildings. This one type of ecosystem could not hold very many different plants and animals. There might not be enough resources left to grow food crops. The air and water might become very dirty. People need trees to help clean the air and water. If the whole earth was a city, there would not be enough trees to clean the air and water.

The earth is not likely going to become one big city. There are many different types of biomes and ecosystems on earth. However, the resources in these ecosystems are not endless. It is important to take care of earth and not destroy ecosystems and habitats.

Conservation is one way people can help ecosystems and habitats to keep being a great place for plants and animals to live. **Conservation** is the act of protecting or restoring a natural environment or ecosystem.

Planting trees is one way to conserve habitats.

Let's look at different ways people can help **conserve** ecosystems.

LAND

Loss of habitat due to buildings and roads is a main factor in endangering plants and animals.

In the United States, certain groups help protect the land. Some groups make sure that companies do not pollute the land.

The National Parks and National Reserves set aside land. Companies are not allowed to build on this land. The habitat is preserved for plants and animals.

Another way to practice conservation is to plant trees. When companies remove trees from forests, they should be replanted.

AIR

Living things need air. Some people have asthma *[AZ-muh]*. Asthma can make it hard for a person to breathe. If a person with asthma lives in a place where the air is polluted, the asthma becomes worse. People and animals can get other sicknesses when the air is dirty.

Air becomes polluted by the smoke from cars and factories. Cars today burn much cleaner smoke than cars in the 1970s. People can practice conservation by driving less. People can also drive cars that **emit** less smoke.

People can become sick when the air is polluted.

WATER

You learned in another Lesson that all living things need water to survive. Many organisms cannot survive in polluted water. In the past, factories in the United States dumped many poisonous chemicals into the water. If a plant takes up polluted water, it can get sick or die. If animals live in or drink polluted water, they can also get sick or die. Today, most companies do not put chemicals into rivers. The water is much cleaner today than it used to be. The water is cleaner because of conservation.

Polluted water harms and kills organisms.

TRASH

One way to help conserve habitats is to clean up trash, even if you didn't make the trash!

You can help conserve water and land resources by not littering. When you litter, trash becomes part of the land or water. Some types of trash have harmful **chemicals** in them. Other types of trash get eaten by animals. Many types of plastics kill birds, sea lions, and turtles every year.

Another way to practice conservation is to make less trash. Reuse any items that you can. Do not buy more things than what you need. Reusing to make less trash leads to the next point: Reduce, Reuse, Recycle.

REDUCE, REUSE, AND RECYCLE

The best thing to do to conserve ecosystems is to use fewer resources or reduce the amount of stuff you buy or use. If you use less paper, then less trees need to be cut down.

One way to reduce is to reuse an item for something else. A used glass pickle jar can be washed and used to store toys. An old box can become a go-cart.

Another way to reduce is to recycle. Electronic equipment like phones and computers should be recycled. Phones contain metals that have to be dug up from the earth. By recycling phones, fewer habitats will be damaged.

Match the words with the descriptions.

9.1) _____ chemical

9.2) _____ conservation

9.3) _____ conserve

9.4) _____ emit

A. a type of matter that can interact with other types of matter

B. to protect from harm

C. to put out

D. the act of protecting or restoring a natural environment or ecosystem

Underline the best answers.

9.5) Driving less or driving cars that emit less smoke is a way to conserve (**land**, **air**, **water**).

9.6) The National Parks and National Reserves preserve habitat for plants and animals. This is an example of conserving (**land**, **air**, **water**).

9.7) Most companies do not put chemicals into rivers like companies in the past did because they want to conserve (**land**, **air**, **water**).

9.8) Planting trees is a way to conserve (**land**, **air**, **water**).

9.9) It is important to conserve (**land**, **air**, **water**) because many organisms cannot survive in it if it is polluted.

9.10) It is important to conserve (**land**, **air**, **water**) because if it is polluted, it can make a person's asthma worse.

9.11) It is important to conserve (**land**, **air**, **water**) because buildings and roads cause loss of habitat.

9.12) When you (**conserve**, **litter**, **emit**), trash becomes part of the land or water.

9.13) It is important not to litter, because some types of trash have harmful (**chemicals**, **smoke**, **animals**) in them.

9.14) It is important not to litter, because some types of trash get (**emitted**, **conserved**, **eaten**) by animals.

Complete the activity.

9.15) Give an example of a way you could conserve in each category.

a. reduce – _____

_____

_____

b. reuse – _____

_____

_____

c. recycle – _____

_____

_____

ACTIVITY
ECOSYSTEM GAME

In this investigation, you will classify, predict, and communicate results. You will make a set of notecards. The notecards include different habitats, organisms, weather, and disasters. These cards will be shuffled and randomly chosen. The back of the cards should be blank and cards should lay facedown so you do not know which card you are drawing.

MATERIALS

- notecards
- markers

PROCEDURE

1 – Make and color four notecards with these habitats on them:

- ocean
- forest
- desert
- swamp

2 – Make and color sixteen animal cards:

- lion
- bear
- wolf
- rabbit
- grasshopper
- salmon
- squirrel
- hawk
- snake
- turtle
- shark
- turkey
- frog
- butterfly
- seagull
- vulture
- lizard
- human

3 – Make and color eight plant cards:

- fruit trees
- cactus
- pine tree
- flowers

- seaweed
- corn
- lily pad
- yam

4 – Make and color four weather cards:

- snow
- rain

- drought
- sunny

5 – Make and color four disaster cards and include four blank cards in this pile:

- pesticide
- earthquake
- flood

- wildfire
- Add the four blank cards to this pile to lessen the chance of a disaster striking.

TO PLAY THE GAME

1 – Shuffle each group of cards facedown so that you do not see which card is which. Place them facedown in their five separate piles.

2 – Draw one card each from the habitat, weather, and disaster piles.

3 – Draw eight animal cards and four plant cards.

4 – Lay your cards faceup in a line in front of you.

5 – Organize the animals into a food chain if possible.

6 – In your journal, go to the next blank page. Title the page "Habitat Game," and write the page number in the upper right. Go to the Table of Contents page, and add the title and page number.

7 – Draw a picture of the food chain in your journal.

8 – Write down in your journal which animals and plants will not survive properly in the habitat you drew.

9 – Predict which animals might be able to adapt to this environment and how. Write your prediction in your journal.

10 – Predict how the weather will affect the plants and animals. Write your prediction in your journal.

11 – Predict the effects of the disaster, if any took place. Write your prediction in your journal.

12 – Reshuffle the piles and redraw two more times. Repeat steps 5, 7, 8, 9, 10, and 11 for each set of cards.

13 – Share the results with your teacher.

GRADING CHECKLIST

❏ All habitat, animal, plant, weather, and disaster cards are made and colored.

❏ Three food chains are drawn in the journal based on what cards are chosen.

❏ Predictions are made for all three situations.

Teacher Check

QUIZ 3

(Each answer, 4 points)

Fill in the blanks using words from the box below.

conservation	chemical	drought	emit
endangered	extinct	fossil	imprint
model	pesticide		

3.01) To_____ is to put out.

3.02) A(n)_____ is an object that shows how something works or what something looks like.

3.03) _____ describes an organism that used to live on earth but whose population completely died out.

3.04) A(n)_____ is the remains of a plant or animal.

3.05) A(n) _____ is used to kill the insects that might eat crops.

3.06) A(n) _____ is a type of matter that can interact with other types of matter.

3.07) _____ is the act of protecting or restoring a natural environment or ecosystem.

3.08) A(n) _____ fossil is a fossil that is the mark of a plant or animal.

3.09) _____ describes an organism whose population is in danger of dying out.

3.010) _____ happens when an area gets less rainfall than normal for a long period of time.

Write L if the example is a way to conserve Land, A for Air, or W for Water.

3.011) _____ Plant trees.

3.012) _____ Stop companies from putting chemicals into rivers.

3.013) _____ Drive less or drive cars that emit less smoke.

3.014) _____ Preserve habitat for plants and animals.

Choose the best answers.

3.015) _____ Which of these is a natural ecosystem change?
   A. drought
   B. using natural resources
   C. pollution
   D. all of these

3.016) _____ Which of these is a man-made ecosystem change?
   A. building   C. catching a ride
   B. pollution   D. all of these

3.017) _____ Which example of pollution can make animals sick?
   A. Factories and cars put smoke into the air.
   B. Ships carrying oil can leak or spill the oil into the ocean.
   C. People can throw their trash on the ground.
   D. all of these

3.018) _____ Why is it bad for an animal to be taken to a new place where it has no predators?
   A. The animal might not survive in the new habitat.
   B. The animal might eat all the plants and leave no food for other animals.
   C. The animal might become food for another animal.
   D. all of these

3.019) _____ Which of these can cause plants and animals to become endangered?
   A. pollution
   B. disappearing habitats
   C. overhunting by humans
   D. all of these

3.020) _____ How are fossils found in rock formed?
   A. An animal dies in mud and the mud turns to rock.
   B. An animal is covered in amber and the amber is surrounded by rock.
   C. An animal is frozen in ice that is replaced by rock.
   D. A fossil forms as the remains of the hard parts of the animal.

3.021) _____ How are endangered animals on the Endangered Species List helped?
   A. They are taken into captivity until they go extinct.
   B. Their fossils are studied because they are extinct.
   C. Their habitats are protected and money is spent to keep them from becoming extinct.
   D. Rules are made to stop them from ever being taken into captivity so that they do not become extinct.

3.022) _____ Some types of trees have seeds that will only germinate if this type of ecosystem change happens.
   A. storm   C. catching a ride
   B. fire   D. drought

3.023) _____ Why is it important not to litter?
   A. Some types of trash have harmful chemicals in them.
   B. Some types of trash can help an animal to take over an ecosystem.
   C. Trash can make a person's asthma worse.
   D. all of these

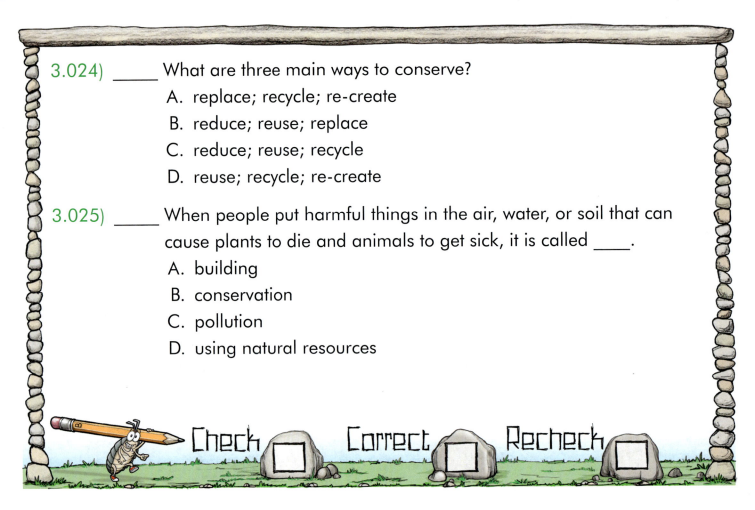

3.024) _____ What are three main ways to conserve?

    A. replace; recycle; re-create

    B. reduce; reuse; replace

    C. reduce; reuse; recycle

    D. reuse; recycle; re-create

3.025) _____ When people put harmful things in the air, water, or soil that can cause plants to die and animals to get sick, it is called _____.

    A. building

    B. conservation

    C. pollution

    D. using natural resources

Check ☐   Correct ☐   Recheck ☐

STOP and prepare for the Unit Practice Test.

- Review the Objectives and the Vocabulary.
- Reread the questions from each Lesson.
- Review the Quizzes.

(Each answer, 2.5 points)
Match the words with the descriptions.

1) _____ organism

2) _____ community

3) _____ habitat

4) _____ population

5) _____ ecosystem

6) _____ biome

7) _____ model

8) _____ extinct

9) _____ endangered

10) _____ conservation

11) _____ fossil

A. a living thing

B. describes an organism whose population is in danger of dying out

C. a place an organism lives in its ecosystem

D. all the different populations living in the same place at the same time

E. the act of protecting or restoring a natural environment or ecosystem

F. all the living and nonliving things that interact in a specific place

G. the remains of a plant or animal

H. an object that shows how something works or what something looks like

I. a group of one kind of organism living in one place at the same time

J. an ecosystem with certain plants, animals, and weather that covers a large area

K. describes an organism that used to live on earth but whose population completely died out

Choose the best answers.

12) _____ Which of these things found in an ecosystem is biotic?

    A. rocks

    B. fish

    C. water

    D. all of these

13) _____ A(n) ___ fossil is a fossil that is the mark of a plant or animal.

    A. imprint

    B. amber

    C. ice

    D. old

14) _____ Which of these things found in an ecosystem is abiotic?

    A. gases

    B. fish

    C. birds

    D. worms

15) _____ Which of these adaptations helps an animal to catch food?

    A. A hawk has feet with talons on them.

    B. A spider makes silk webbing.

    C. Wolves hunt in packs.

    D. all of these

16) _____ Which of these adaptations helps an animal to get water?

    A. A thorny devil lizard has grooves on its back.

    B. Dolphins live together in pods.

    C. Ducks have webbed feet.

    D. all of these

17) _____ All of the sunflowers in the field in the picture make up a(n) _____.

A. habitat
B. organism
C. population
D. community

18) _____ A special process that plants use to make their own food is ___.
A. respiration
B. mimicry
C. predation
D. photosynthesis

19) _____ The ability to blend into the background of the environment is called ___.
A. camouflage
B. mimicry
C. predation
D. photosynthesis

20) _____ The ability of an animal or plant to look like a more dangerous predator is ____.
A. camouflage
B. mimicry
C. predation
D. photosynthesis

21) _____ The grasshoppers in this food chain are ____.

A. consumers
B. predators
C. prey
D. A and C

22) _____ How does camouflage help a predator to get food?

    A. It helps the predator to sneak up to the prey.

    B. It makes it harder for the predator to find the prey.

    C. It helps the predator to look more dangerous than it is.

    D. It allows the predator to keep the prey from moving.

23) _____ Which of these is an adaptation that helps a predator get food?

    A. poison   C. mimicry

    B. smell   D. all of these

24) _____ Which of these is an adaptation that helps a prey not get eaten?

    A. camouflage   C. smell

    B. spikes and thorns   D. all of these

25) _____ Planting trees is a way to conserve ___.

    A. land   B. air   C. water

26) _____ Stopping companies from putting chemicals into rivers is a way to conserve ___.

    A. land   B. air   C. water

27) _____ Driving less and walking more is a way to conserve ____.

    A. land   B. air   C. water

28) _____ Preserving habitat for plants and animals is a way to conserve ____.

    A. land   B. air   C. water

29) _____ Which of these is a natural ecosystem change?

    A. drought

    B. using natural resources

    C. pollution

    D. all of these

30) _____ Which of these is a man-made ecosystem change?

    A. building   C. catching a ride

    B. pollution   D. all of these

31) _____ Why is it bad for an animal to be taken to a new place where it has no predators?

    A. The animal might not survive in the new habitat.

    B. The animal might eat all the plants and leave no food for other animals.

    C. The animal might become food for another animal.

    D. all of these

32) _____ How are endangered animals on the Endangered Species List helped?

    A. They are taken into captivity until they go extinct.

    B. Their fossils are studied because they are extinct.

    C. Their habitats are protected and money is spent to keep them from becoming extinct.

    D. Rules are made to stop them from ever being taken into captivity so that they do not become extinct.

Match the names with the parts of photosynthesis.

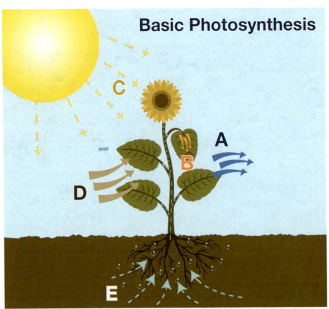

Basic Photosynthesis

33) _____ water

34) _____ carbon dioxide

35) _____ oxygen

36) _____ sugar

37) _____ light

88

Use the food web to choose the best answers.

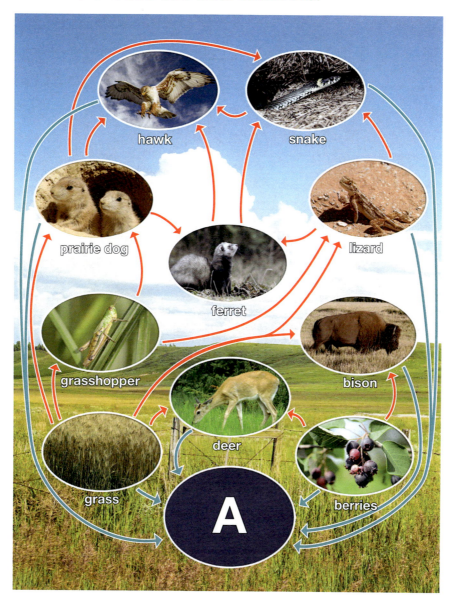

38) _____ What belongs in the box labeled A?

A. producer   B. consumer   C. decomposer   D. herbivore

39) _____ What type of organism is the snake?

A. producer   B. consumer   C. decomposer   D. herbivore

40) _____ What type of organism is the grass?

A. producer   B. consumer   C. decomposer   D. herbivore

You must now prepare for the Unit Test.

- Review the Objectives and the Vocabulary.
- Reread the questions from each Lesson.
- Review and study the Quizzes and Unit Practice Test.

When you are ready, turn in your Unit and request your Unit Test.